The Mini Sabb a Field Guide

By

Kimberly Buyze

Table of Contents

The Purpose of the Field Guide

Many Pastors and Professors take sabbaticals, but can the layperson take one too? What are the benefits of a sabbatical, and frankly, what does one do on a sabbatical? I am a software engineer who felt led to take a mini, four-week sabbatical after 10 years at my company. My goal was rest and restoration, along with spiritual growth and renewal. I wanted to reflect on the past ten years of my career while thinking about the next ten and where they might take me. As I looked for guides on sabbatical, I found several resources for preparing for a sabbatical and others for re-integrating into normal life, but none explaining how to "do" a sabbatical. While there are no hard-set rules for sabbaticals, this field guide gives some ideas on how to use the time intentionally and experience rest and restoration without further burnout or wasting time. I believe everyone can benefit from a sabbatical. Teachers, students, or folks in between jobs may be able to naturally fit in a sabbatical, but there are no boundaries from anyone taking one for a short length of time.

Why Sabbatical?

First, I would like to focus on why it is important to take a sabbatical.

Sabbath

The Bible gives descriptions of times when we are to work and rest. Exodus 23:10 - 13 notes:

10 For six years you are to sow your fields and harvest the crops, 11 but during the seventh year let the land lie unplowed and unused. Then the poor among your people may get food from it, and the wild animals may eat what they leave. Do the same with your vineyard and your olive grove.

12 Six days do your work, but on the seventh day do not work, so that your ox and your donkey may rest, and so that the slave born in your household, and the alien as well, may be refreshed.

13 Be careful to do everything I have said to you. Do not invoke the names of other gods; do not let them be heard on your lips. [1]

From this, we see two cycles of work and rest. The first is after six years of work to allow the land to rest. The second is weekly rest, often referred to as sabbath. I have a friend who mentioned she often sees seasons in her life come in seven-year cycles. She stays in one place for seven years and then feels moved to another. My life is not quite so ordered, but I have seen much of the same thing.

Pilgrimage

Psalm 84:5-7 reads:

5 Blessed are those whose strength is in you,

Who have set their hearts on pilgrimage.

6 As they pass through the Valley of Baka,

they make it a place of springs;

the autumn rains also cover it with pools.

7 They go from strength to strength,

till each appears before God in Zion. [1]

Many people take actual pilgrimages (Jerusalem in Israel or the Camino de Santiago in Spain are popular). But I believe a sabbatical is also a form of pilgrimage. Many of the days during my sabbatical, I thought of the quote from Frozen 2, "The Enchanted Forest will bring transformation." [2] As we travel with God through a sabbatical, we will experience His transforming work.

Sacrifice

Jesus shared with his disciples in John 12:24-26:

24 I tell you the truth, unless a kernel of wheat falls to the ground and dies, it remains only a single seed. But if it dies, it produces many seeds.

25 The man who loves his life will lose it, while the man who hates his life in this world will keep it for eternal life.

26 Whoever serves me must follow me; and where I am, my servant also will be. My Father will honor the one who serves me. [1]

As I was preparing for my sabbatical, I came across this verse in the famous 23 Psalm:

2 He makes me lie down in green pastures, he leads me beside quiet waters. [1]

When I read it, I laughed. Why does God have to make us lie down in green pastures? Wouldn't we do it naturally? It seems rest is not at the forethought of our minds and is something we must intentionally do. When we are on sabbatical, we may think, "What have I done?" or is this "Too much?" We may worry about our jobs or the other things we should be doing. Stick with the process; the transformation will come, and the fruit will be seen.

Notes on Sabbatical

Before we dive into a proposed format for the sabbatical, I would like to highlight some surprises that hit me during my sabbatical and just good advice I received in preparation for my sabbatical.

Warfare

The enemy does not want us to take a sabbatical. He does not want us to draw nearer to Jesus; He does not want us to take time for rest and restoration.

1 Peter 5:8-9 notes:

8 Be self-controlled and alert. Your enemy the devil prowls around like a roaring lion looking for someone to devour.

9 Resist him, standing firm in the faith, because you know that your brothers throughout the world is undergoing the same kind of sufferings. [1]

I often pray whatever is attacking me will flee in the name of Jesus. It is also good to have prayer partners during your sabbatical (see Team section).

Before I started my sabbatical, I was hit with three significant relational challenges. I don't believe these came by accident, but were attacks of the enemy. I also experienced mild panic attacks and felt a "darkness of the soul" during the first week of my sabbatical. It is good to remember 1 John 5:11-12 during these times of attack:

11 And this is the testimony: God has given us eternal life, and this life is in his Son. 12 Whoever has the Son has life; whoever does not have the Son of God does not have life.

It is also good to remember whatever you are giving up (time at work, relationships, etc.) will still be there when you are finished and that they are in God's hands.

Grief

It is not uncommon to experience sessions of grief during a sabbatical. I encourage us to lean into these times of grief and let the tears fall. Be open and honest with Jesus about your grief and what you are feeling. There is always a sweet feeling of peace and comfort after I have allowed myself to grieve. I have talked with others who have done sabbaticals and experienced grief as well.

Matthew 5:3:

Blessed are the poor in spirit, for theirs is the kingdom

of heaven. [1]

Team

In preparation for my sabbatical, I read the book "Desperate Rest: Restoring Your Soul Through Sabbatical" by Laura Demetrician, LMFT. Laura suggested preparing a team that can pray for you and check in with you during your sabbatical. [3] I believe my team was essential in helping me process what God was doing in and through me during my sabbatical, as well as providing prayer covering.

Connection

During a sabbatical, it is important to stay connected with other human beings. A few weeks before my sabbatical, God reminded me of a rule of life I had created not to go twenty-four hours without seeing another person. A friend at church confirmed this rule by commenting I may want to make sure I have connections with other people and not become isolated. I challenged myself to do my reflection time in public places (a neighborhood clubhouse, a Christian organization that had prayer rooms for reservation, and coffee shops). This way, I made sure I had contact with other people. This kept me connected to the outside world and helped me control my thoughts. I would also say be careful how much tweeting and texting is done during a sabbatical. You may have a thought that seems ingenious during the sabbatical but is embarrassing afterward. Don't turn into a bible banger; let thoughts and scriptures marinate, and feel free to share them after your sabbatical time. I try to leave nothing unsaid but recognize I don't have to say everything at once.

Detachment

Another good nugget from Laura Demetrician was to consider how many commitments you want to keep during a sabbatical. [3] I decided to keep meeting with my small group but step down from church volunteering and reading technical books that had to do with my vocation. This is a precious time and will need to be guarded. I also found I had a fairly low tolerance for "drama" and decided to put some relationships on hold.

Rhythm

In 2005, I moved to Brazil for a yearlong internship. When I first arrived, I felt very disoriented. One of the interns who had been in Brazil longer than me said I would feel better once I started working and got into a routine. I believe it is the same with a sabbatical. Plan a "rhythm" for your sabbatical to help keep you grounded. My "rhythm" was to do spiritual practices Monday through Friday and have "normal" weekends. I put together a rough schedule (see sample below) for what I wanted to do each day. I believe a rhythm also helps us keep intentional with our sabbatical and not drift. It also helps others pray. Note you don't have to stick to your schedule or do everything on it, but it will help keep you focused.

Example Sabbatical Rough Schedule

7–9 a.m. Prayer Retreat/Foot Washing

9–10 a.m. Rest/Language Training

10–12 a.m. Hike/Prayer Walk

12–2 p.m. Lunch/Rest

2–4 p.m. Reflection Time

4–6 p.m. Dinner/Rest

6–7 p.m. Bible Study/Writing

7–8 p.m. Strength Training

Theme

One of the friends I reached out to for advice on "how to sabbatical" recommended praying for a theme for the sabbatical. My theme was "Fruits of the Spirit," and the theme helped focus my time on the Bible.

Format

In this section, we will focus on different activities you may want to do during your sabbatical. This will be planned into your Monday through Friday rhythm. I will highlight different areas you may want to incorporate into your sabbatical. My sabbatical was short (1 month), so I decided not to do any vocational activities. For longer sabbaticals, you may want to try to learn something new related to your vocation, but be careful to time box the amount of time spent on vocational activities so they don't become overwhelming.

Prayer

I believe prayer is essential in everything we do. In her "Lit" study, Beth Moore noted she was concerned about how much American Christians are praying. She recommended people in ministry pray at least two hours a day. Later in this guide is a "Prayer Retreat." I have found prayer retreats helpful in guiding my prayer time. Beth Moore also noted the importance of interceding for others and not just praying for self needs. I felt convicted in this as I realized my intercessory prayer time was getting shorter and shorter. [4]

Play

Chapter one of John Eldredge's book "Beautiful Outlaw" is titled "The Playfulness of God and the Poison of Religion." [5] When my family goes rocking, climbing, and rappelling in the breathtaking Colorado Rockies, we often take time to thank God for his beautiful creation and invite him into our play. A few months prior to my sabbatical, I felt God was asking me to invite him into my "play" and fun activities. I believe God loves it when we play and experience his creation.

Jesus said in Matthew 19:14:

"Let the little children come to me, and do not hinder them, for the kingdom of heaven belongs to such as these." [1]

He also noted in Luke 18:17:

"I tell you the truth, anyone who will not receive the kingdom of God like a little child will never enter it." [1]

For my sabbatical, I chose to engage in short hikes for my "play," but there are numerous other activities that can be done as well.

Reflection

Pastor Mark Balmer once said, "If nothing changes, nothing changes." [6] Reflecting on our past helps us see patterns that might need change and helps us prepare for the future. It also gives us a chance to celebrate our successes. This guide offers questions for reflection. Take a moment to ponder each question and be open and honest in your response. You will be surprised by what you may find.

Study

Proverbs 3:7-8 tells us:

7 Do not be wise in your own eyes;

fear the Lord and shun evil.

8 This will bring health to your body

and nourishment to your bones. [1]

Taking time to read through a book or participate in a video study during a sabbatical can be very life-giving. I chose to do the "Lit" and "Fulfill your Ministry" studies by Beth Moore. There are many authors and teachers who can help you grow in your walk with the Lord.

Sacrament

Wikipedia defines a sacrament as "a Christian rite that is recognized as being particularly important and significant. There are various views on the existence, number, and meaning of such rites. Many Christians consider the sacraments to be a visible symbol of the reality of God, as well as a channel for God's grace. Many denominations, including the Roman Catholic, Lutheran, Presbyterian Anglican, Methodist, and Reformed, hold to the definition of sacrament formulated by Augustine of Hippo: an outward sign of an inward grace that has been instituted by Jesus Christ. Sacraments signify God's grace in a way that is outwardly observable to the participant." [7]

Sacraments are a tangible way to help us remember Jesus and draw us closer to God. Some sacraments we can participate in during the sabbatical include:

Communion

Foot Washing

Remembering Baptism by the Sprinkling of Water

Fasting a Meal

Anointing with Oil

Enjoying a "Simple" Meal of Bread and Water

Exercise

1 Corinthians 6:19-20 reminds us:

19 Do you not know that your body is a temple of the Holy Spirit, who is in you, whom you have received from God? You are not your own;

20 you were bought at a price. Therefore, honor God with your body. [1]

Getting fresh air and having an elevated heart rate is good for the mind, body, and soul. It also helps us connect with others and see God's creation. I chose to do hikes as well as strength training during my sabbatical, but there are a myriad of ways to move around and exercise.

Rest

Rest is a biblical concept. The Bible tells us:

8 "Remember the Sabbath day by keeping it holy.

9 Six days you shall labor and do all your work,

10 but the seventh day is a Sabbath to the Lord your God. On it you shall not do any work, neither you, nor your son or daughter, nor your manservant, or maidservant, nor your animals, nor the alien within your gates.

11 For in six days the Lord made the heavens and the earth, the sea, and all that is in them, but he rested on the seventh day. Therefore, the Lord blessed the Sabbath day and made it holy." [1] (Exodus 20:8-11)

There is also the concept of entering God's rest. Hebrews 4:1-3a tells us:

1Therefore, since the promise of entering his rest still stands, let us be careful that none of you be found to have fallen short of it.

2 For we also have had the gospel preached to us, just as they did; but the message they heard was of no value to them, because those who heard it did not combine it with faith.

3 Now we who have believed enter that rest, just as God has said:

"So I declared on oath in my anger,

'They shall never enter my rest.'" [1]

Finally, after periods of battle, the Israelites experienced rest from their enemies. Joshua 21:44 tells us:

The Lord gave them rest on every side, just as he had sworn to their forefathers. Not one of their enemies withstood them; the Lord handed all their enemies over to them. [1]

It may be fun as a thought exercise to write down what rest on every side looks like and pray it over our lives.

As we plan our sabbatical rhythm, we should add periods of rest so as not to get overwhelmed. Rest, for me, looks like reading a lighthearted book, coloring, or taking a nap. Feel free to discover what rest looks like for you.

Be careful how much screen time is spent on a sabbatical. John Eldridge wrote, "Relief is momentary; it's checking out, numbing, sedating yourself. Television is a relief. Eating a bag of cookies is a relief. Tequila is a relief. And let's be honest—relief is what we reach for because it's immediate and usually within our grasp. Most of us turn there when what we really need is restoration.

Nature heals; nature restores. Think of sitting on the beach watching the waves roll in at sunset and compare it to turning on the tube and vegging in front of Narcos or Fear the Walking Dead. The experiences could not be further apart. Remember how you feel sitting by a small brook, listening to its little musical songs, and contrast that to an hour of HALO. Video games offer relief; nature offers restoration." [8]

Work

We were created to work. In Genesis 2:15 we see:

The Lord God took the man and put him in the Garden of Eden to work it and take care of it. [1]

I don't believe sabbaticals should be spent vegging out on the couch and watching TV. Setting a rhythm will help keep the sabbatical focused and intentional. I choose to follow a loose schedule Monday through Friday and let weekends be more ad-hoc. While there is work being done during the sabbatical, stay in step with the Spirit. Feel free to stop and take a nap or drop something from your rhythm if it is too overwhelming. The goal is rest and restoration.

Planning

Often, people will use sabbatical time to set goals and plan for the future. A friend told me his organization uses the sabbatical for members to check in with God to see if they are where they are supposed to be. I worked for fifteen years in corporate America before I began setting goals different from advancing in my career and then retirement. I don't believe a sabbatical has to be used solely for planning. The rest and restoration in and of itself is healthy but listen for what God may be saying about your future and where He may lead you. A quote I love from the book Experiencing God is "Reality 4: God speaks by the Holy Spirit through the Bible, prayer, circumstances, and the church to reveal Himself, His purposes, and His ways." [9]

Creative

The creative side of the sabbatical is my favorite part. Working as an engineer, I often get to add creativity to my designs, but most of my day is highly technical. Add some time to create into your sabbatical. I enjoyed spending more time playing guitar, cooking nice meals, and doing some writing. Some other things you could do are drawing or painting, pottery, taking a dance class or spending time on a musical instrument. The sky is the limit!

Reintegration

Toward the end of my sabbatical, I planned five days with a friend in Kauai to enjoy a new place and begin to get used to making decisions again. It was also a great way to celebrate! Begin to think about how you will tackle life's challenges and how you will add items (work, relationships, etc) back into your life.

Mini Sabbatical Journaling Pages

This next section provides journaling space and guidelines for each day of your sabbatical. Feel free to use all or portions of the journaling pages as guidelines during your sabbatical journey. The journaling section covers three weeks, Monday through Friday, but can be adapted to any length of time. The prayer retreat format is a combination of prayer retreats I have taken with Adventures in Missions and with "Miss Peggy" Alberda of Restored to Glory Dance. The prayer format is taken from the **A**doration, **C**onfession, **T**hanksgiving, **S**upplication/Intercession (**ACTS**) prayer guideline with searching, listening and application added to the end. Plan 1.5 to 3 hours for the prayer retreat and around 30 minutes for the reflection. The prayer retreat and reflection do not need to be done together.

Week 1, day 1: Love part one

Visit your schedule to re-familiarize yourself with the rhythm you have set up. An example can be found in the "Notes on Sabbatical" section. See how you may want to alter it for today.

Prayer retreat (1.5 - 3 hours)

Adoration

Some people enjoy coming into God's presence through music. Sometimes, I pull out my guitar or listen to YouTube videos. One could also read a prayer or moving bible passage. I try to listen to the Holy Spirit and what songs or passages God is putting in my heart to draw near to Him. Toward the end of the time, read Psalm 19. Consider how God has impacted your life. Consider the greatness of God. Feel free to write any thoughts in the space below.

_____

_____

_____

_____

_____

_____

_____

_____

_____

Confession

Read Romans 10, focusing on verse 9. Seek the Lord. What needs to be confessed before Him? Attitudes, thoughts, bad language? Read through the 10 commandments. Cry aloud to the Lord.

Thanksgiving

Take time to journal things you are thankful for. Write about your life, health, etc. Write about anything in the reflection sections you feel grateful for. Read Nehemiah 8, focusing on verse 10.

Intercession

Begin to pray for others. This section is key as we can often get self-focused. How can you pray for immediate family, extended family, pastors, co-workers, nations in need, your country, people in leadership, and non-profits? Let the words flow. Read Ephesians 1, focusing on verse 18.

_____

_____

_____

_____

_____

_____

_____

_____

_____

_____

_____

_____

_____

Searching

Humbly begin praying for yourself. Make a list of needs and give them one at a time to God, expressing your heart. What events are coming up in your life? What career goals do you have? What life goals do you have? Do any words stick out from your confession time? God is listening. Read Job 38 or a passage from any biblical character that inspires you.

_____

_____

_____

_____

_____

_____

_____

_____

_____

_____

_____

_____

_____

Listening

Begin listening to God for passages of scripture/answers to prayer. If nothing sticks out, feel free to use the reflection scriptures below.

Scriptures for reflection on Love:

Romans 12:9

Romans 13:8a

_____

_____

_____

_____

_____

_____

_____

_____

_____

_____

_____

_____

Application

Pray for strength to carry out anything God has revealed to you during the prayer retreat today. Read Psalm 46.

Reflection (30 minutes - 1 hour) separate from prayer retreat

Achievements

1. What achievement are you most proud of in the past 10 years?

_____

_____

_____

_____

_____

_____

_____

_____

_____

_____

_____

_____

_____

2. What event made you the most nervous in the past 10 years?

_____

_____

_____

_____

_____

_____

_____

_____

_____

_____

_____

_____

_____

_____

_____

_____

_____

_____

3. What hurtle did you overcome in the past 10 years?

_____

_____

_____

_____

_____

_____

_____

_____

_____

_____

_____

_____

_____

_____

_____

_____

4. What awards/accolades were you given in the past 10 years? Why?

_____

_____

_____

_____

_____

_____

_____

_____

_____

_____

_____

_____

_____

_____

_____

_____

_____

5. What goals did you accomplish in the past 10 years?

_____

_____

_____

_____

_____

_____

_____

_____

_____

_____

_____

_____

_____

_____

_____

_____

Journaling (10 minutes)

What stuck out to you today?

_____

_____

_____

_____

_____

_____

_____

_____

_____

_____

_____

_____

_____

_____

_____

Week 1, day 2: Love part two

Visit your schedule to re-familiarize yourself with the rhythm you have set up. An example can be found in the "Notes on Sabbatical" section. See how you may want to alter it for today.

Prayer retreat (1.5 - 3 hours)

Adoration

Some people enjoy coming into God's presence through music. Sometimes, I pull out my guitar or listen to YouTube videos. One could also read a prayer or moving bible passage. I try to listen to the Holy Spirit and what songs or passages God is putting in my heart to draw near to Him. Toward the end of the time, read Psalm 19. Consider how God has impacted your life. Consider the greatness of God. Feel free to write any thoughts in the space below.

_____

_____

_____

_____

_____

_____

_____

_____

_____

_____

Confession

Read Romans 10, focusing on verse 9. Seek the Lord. What needs to be confessed before Him? Attitudes, thoughts, bad language? Read through the 10 commandments. Cry aloud to the Lord.

_____

_____

_____

_____

_____

_____

_____

_____

_____

_____

_____

_____

_____

_____

Thanksgiving

Take time to journal things you are thankful for. Write about your life, health, etc. Write about anything in the reflection sections you feel grateful for. Read Nehemiah 8, focusing on verse 10.

Intercession

Begin to pray for others. This section is key as we can often get self-focused. How can you pray for immediate family, extended family, pastors, co-workers, nations in need, your country, people in leadership, and non-profits? Let the words flow. Read Ephesians 1, focusing on verse 18.

Searching

Humbly begin praying for yourself. Make a list of needs and give them one at a time to God, expressing your heart. What events are coming up in your life? What career goals do you have? What life goals do you have? Do any words stick out from your confession time? God is listening. Read Job 38 or a passage from any biblical character that inspires you.

_____

_____

_____

_____

_____

_____

_____

_____

_____

_____

_____

_____

Listening

Begin listening to God for passages of scripture/answers to prayer. If nothing sticks out, feel free to use the reflection scriptures below.

Scriptures for reflection on Love:

Galatians 5:13-15

1 Corinthians 13

_____

_____

_____

_____

_____

_____

_____

_____

_____

_____

_____

Pray for strength to carry out anything God has revealed to you during the prayer retreat today. Read Psalm 46.

Reflection (30 minutes - 1 hour) separate from prayer retreat

Hopes

1. What do you want to accomplish in the next 10 years' career-wise?

_____

_____

_____

_____

_____

_____

_____

_____

_____

_____

_____

_____

_____

_____

_____

_____

_____

2. What do you want to accomplish in the next 10 years' relationship-wise?

_____

_____

_____

_____

_____

_____

_____

_____

_____

_____

_____

_____

_____

_____

_____

3. Where do you want to live in 10 years?

_____

_____

_____

_____

_____

_____

_____

_____

_____

_____

_____

_____

_____

_____

_____

_____

_____

4. What is your dream job?

_____

_____

_____

_____

_____

_____

_____

_____

_____

_____

_____

_____

_____

_____

_____

_____

_____

5. What was your dream job as a child, and how is it different now?

_____

_____

_____

_____

_____

_____

_____

_____

_____

_____

_____

_____

_____

_____

_____

Journaling (10 minutes)

What stuck out to you today?

_____

_____

_____

_____

_____

_____

_____

_____

_____

_____

_____

_____

_____

_____

Week 1, day 3: Joy part one

Visit your schedule to re-familiarize yourself with the rhythm you have set up. An example can be found in the "Notes on Sabbatical" section. See how you may want to alter it for today.

Prayer retreat (1.5 - 3 hours)

Adoration

Some people enjoy coming into God's presence through music. Sometimes, I pull out my guitar or listen to YouTube videos. One could also read a prayer or moving bible passage. I try to listen to the Holy Spirit and what songs or passages God is putting in my heart to draw near to Him. Toward the end of the time, read Psalm 19. Consider how God has impacted your life. Consider the greatness of God. Feel free to write any thoughts in the space below.

Confession

Read Romans 10, focusing on verse 9. Seek the Lord. What needs to be confessed before Him? Attitudes, thoughts, bad language? Read through the 10 commandments. Cry aloud to the Lord.

_____

_____

_____

_____

_____

_____

_____

_____

_____

_____

_____

_____

Thanksgiving

Take time to journal things you are thankful for. Write about your life, health, etc. Write about anything in the reflection sections you feel grateful for. Read Nehemiah 8, focusing on verse 10.

Intercession

Begin to pray for others. This section is key as we can often get self-focused. How can you pray for immediate family, extended family, pastors, co-workers, nations in need, your country, people in leadership, and non-profits? Let the words flow. Read Ephesians 1, focusing on verse 18.

Searching

Humbly begin praying for yourself. Make a list of needs and give them one at a time to God, expressing your heart. What events are coming up in your life? What career goals do you have? What life goals do you have? Do any words stick out from your confession time? God is listening. Read Job 38 or a passage from any biblical character that inspires you.

Begin listening to God for passages of scripture/answers to prayer. If nothing sticks out, feel free to use the reflection scriptures below.

Scriptures for reflection on Joy:

Philippians 4:4

Psalm 51:8

Psalm 126

Psalm 5:11-12

Application

Pray for strength to carry out anything God has revealed to you during the prayer retreat today. Read Psalm 46.

Reflection (30 minutes - 1 hour) separate from prayer retreat

Hopes

1. What hopes do you have for your family?

_____

_____

_____

_____

_____

_____

_____

_____

_____

_____

_____

_____

_____

_____

2. What hopes do you have for close friends?

3. What hopes do you have for close co-workers/collogues?

_____

_____

_____

_____

_____

_____

_____

_____

_____

_____

_____

_____

_____

_____

_____

_____

4. What hopes do you have for your physical body? Emotional state?

_____

_____

_____

_____

_____

_____

_____

_____

_____

_____

_____

_____

_____

_____

_____

_____

_____

_____

5. What steps do you see yourself taking to accomplish these hopes?

Journaling (10 minutes)

What stuck out to you today?

_____

_____

_____

_____

_____

_____

_____

_____

_____

_____

_____

_____

Week 1, day 4: Joy part two

Visit your schedule to re-familiarize yourself with the rhythm you have set up. An example can be found in the "Notes on Sabbatical" section. See how you may want to alter it for today.

Prayer retreat (1.5 - 3 hours)

Adoration

Some people enjoy coming into God's presence through music. Sometimes, I pull out my guitar or listen to YouTube videos. One could also read a prayer or moving bible passage. I try to listen to the Holy Spirit and what songs or passages God is putting in my heart to draw near to Him. Toward the end of the time, read Psalm 19. Consider how God has impacted your life. Consider the greatness of God. Feel free to write any thoughts in the space below.

_____

_____

_____

_____

_____

_____

_____

_____

_____

_____

Confession

Read Romans 10, focusing on verse 9. Seek the Lord. What needs to be confessed before Him? Attitudes, thoughts, bad language? Read through the 10 commandments. Cry aloud to the Lord.

_____

_____

_____

_____

_____

_____

_____

_____

_____

_____

_____

_____

_____

Thanksgiving

Take time to journal things you are thankful for. Write about your life, health, etc. Write about anything in the reflection sections you feel grateful for. Read Nehemiah 8, focusing on verse 10.

_____

_____

_____

_____

_____

_____

_____

_____

_____

_____

_____

_____

_____

_____

Intercession

Begin to pray for others. This section is key as we can often get self-focused. How can you pray for immediate family, extended family, pastors, co-workers, nations in need, your country, people in leadership, and non-profits? Let the words flow. Read Ephesians 1, focusing on verse 18.

_____

_____

_____

_____

_____

_____

_____

_____

_____

_____

_____

_____

_____

Searching

Humbly begin praying for yourself. Make a list of needs and give them one at a time to God, expressing your heart. What events are coming up in your life? What career goals do you have? What life goals do you have? Do any words stick out from your confession time? God is listening. Read Job 38 or a passage from any biblical character that inspires you.

_____

_____

_____

_____

_____

_____

_____

_____

_____

_____

_____

_____

Listening

Begin listening to God for passages of scripture/answers to prayer. If nothing sticks out, feel free to use the reflection scriptures below.

Scriptures for reflection on Joy:

Psalm 4:7

Ezra 6:22

2 Corinthians 7:4

1 Kings 8:66

1 Chronicles 29:17

Psalm 16:11

Psalm 21:1

Application

Pray for strength to carry out anything God has revealed to you during the prayer retreat today. Read Psalm 46.

Reflection (30 minutes - 1 hour) separate from prayer retreat

Dreams

1. What travel plans do you have for the next 10 years?

2. Have you thought about retirement?

_____

_____

_____

_____

_____

_____

_____

_____

_____

_____

_____

_____

_____

_____

_____

_____

_____

3. Do you have any career goals?

_____

_____

_____

_____

_____

_____

_____

_____

_____

_____

_____

_____

_____

_____

_____

_____

_____

4. What is the craziest dream you have ever had?

_____

_____

_____

_____

_____

_____

_____

_____

_____

_____

_____

_____

_____

_____

_____

5. Are you a dreamer or more practical?

_____

_____

_____

_____

_____

_____

_____

_____

_____

_____

_____

_____

_____

_____

_____

_____

Journaling (10 minutes)

What stuck out to you today?

_____

_____

_____

_____

_____

_____

_____

_____

_____

_____

_____

_____

_____

_____

_____

Week 1 day 5: Peace part one

Visit your schedule to re-familiarize yourself with the rhythm you have set up. An example can be found in the "Notes on Sabbatical" section. See how you may want to alter it for today.

Prayer retreat (1.5 - 3 hours)

Adoration

Some people enjoy coming into God's presence through music. Sometimes I pull out my guitar or listen to YouTube videos. One could also read a prayer or moving bible passage. I try to listen to the Holy Spirit and what songs or passages God is putting in my heart to draw near to Him. Toward the end of the time, read Psalm 19. Consider how God has impacted your life. Consider the greatness of God. Feel free to write any thoughts in the space below.

_____

_____

_____

_____

_____

_____

_____

_____

_____

_____

Confession

Read Romans 10, focusing on verse 9. Seek the Lord. What needs to be confessed before Him? Attitudes, thoughts, bad language? Read through the 10 commandments. Cry aloud to the Lord.

_____

_____

_____

_____

_____

_____

_____

_____

_____

_____

_____

_____

_____

Thanksgiving

Take time to journal things you are thankful for. Write about your life, health, etc. Write about anything in the reflection sections you feel grateful for. Read Nehemiah 8, focusing on verse 10.

Intercession

Begin to pray for others. This section is key as we can often get self-focused. How can you pray for immediate family, extended family, pastors, co-workers, nations in need, your country, people in leadership, and non-profits? Let the words flow. Read Ephesians 1, focusing on verse 18.

_____

_____

_____

_____

_____

_____

_____

_____

_____

_____

_____

_____

Searching

Humbly begin praying for yourself. Make a list of needs and give them one at a time to God, expressing your heart. What events are coming up in your life? What career goals do you have? What life goals do you have? Do any words stick out from your confession time? God is listening. Read Job 38 or a passage from any biblical character that inspires you.

_____

_____

_____

_____

_____

_____

_____

_____

_____

_____

_____

_____

Listening

Begin listening to God for passages of scripture/answers to prayer. If nothing sticks out, feel free to use the reflection scriptures below.

Scriptures for reflection on Peace:

Philippians 4:7,9

_____

_____

_____

_____

_____

_____

_____

_____

_____

_____

_____

_____

_____

Application

Pray for strength to carry out anything God has revealed to you during the prayer retreat today. Read Psalm 46.

Reflection (30 minutes - 1 hour) separate from prayer retreat

Dreams

1. Take some time to write out… (they can be simple)

A. 1-month goals

_____

_____

_____

_____

_____

_____

_____

_____

_____

_____

_____

_____

_____

_____

_____

B. 1-year goals

_____

_____

_____

_____

_____

_____

_____

_____

_____

_____

_____

_____

_____

_____

_____

_____

_____

C. 5-year goals

_____

_____

_____

_____

_____

_____

_____

_____

_____

_____

_____

_____

_____

_____

_____

_____

D. 10-year goals

Journaling (10 minutes)

What stuck out to you today?

_____

_____

_____

_____

_____

_____

_____

_____

_____

_____

_____

_____

_____

_____

Week 2, day 1: Peace part two

Visit your schedule to re-familiarize yourself with the rhythm you have set up. An example can be found in the "Notes on Sabbatical" section. See how you may want to alter it for today.

Prayer retreat (1.5 - 3 hours)

Adoration

Some people enjoy coming into God's presence through music. Sometimes, I pull out my guitar or listen to YouTube videos. One could also read a prayer or moving bible passage. I try to listen to the Holy Spirit and what songs or passages God is putting in my heart to draw near to Him. Toward the end of the time, read Psalm 19. Consider how God has impacted your life. Consider the greatness of God. Feel free to write any thoughts in the space below.

Confession

Read Romans 10, focusing on verse 9. Seek the Lord. What needs to be confessed before Him? Attitudes, thoughts, bad language? Read through the 10 commandments. Cry aloud to the Lord.

_____

_____

_____

_____

_____

_____

_____

_____

_____

_____

_____

_____

_____

_____

_____

Thanksgiving

Take time to journal things you are thankful for. Write about your life, health, etc. Write about anything in the reflection sections you feel grateful for. Read Nehemiah 8, focusing on verse 10.

Intercession

Begin to pray for others. This section is key as we can often get self-focused. How can you pray for immediate family, extended family, pastors, co-workers, nations in need, your country, people in leadership, and non-profits? Let the words flow. Read Ephesians 1, focusing on verse 18.

Searching

Humbly begin praying for yourself. Make a list of needs and give them one at a time to God, expressing your heart. What events are coming up in your life? What career goals do you have? What life goals do you have? Do any words stick out from your confession time? God is listening. Read Job 38 or a passage from any biblical character that inspires you.

_____

_____

_____

_____

_____

_____

_____

_____

_____

_____

_____

_____

Listening

Begin listening to God for passages of scripture/answers to prayer. If nothing sticks out, feel free to use the reflection scriptures below.

Scriptures for reflection on Peace:

Psalm 46:1-2

2 Timothy 1:7

1 Peter 5:10

Application

Pray for strength to carry out anything God has revealed to you during the prayer retreat today. Read Psalm 46.

Reflection (30 minutes - 1 hour) separate from prayer retreat

Past

1.   How have you cared for yourself in the past 10 years?

_____

_____

_____

_____

_____

_____

_____

_____

_____

_____

_____

_____

_____

_____

_____

_____

_____

_____

2. What goals have you not completed in the past 10 years?

3. What have you missed out on in the past 10 years?

4. Of the things you missed out on, what can you redeem or re-due? (be creative)

_____

_____

_____

_____

_____

_____

_____

_____

_____

_____

_____

_____

5. Where have you given the enemy a stronghold?

_____

_____

_____

_____

_____

_____

_____

_____

_____

_____

_____

_____

_____

_____

_____

Journaling (10 minutes)

What stuck out to you today?

Week 2, day 2: Patience

Visit your schedule to re-familiarize yourself with the rhythm you have set up. An example can be found in the "Notes on Sabbatical" section. See how you may want to alter it for today.

_____

_____

_____

_____

_____

_____

_____

_____

_____

_____

_____

Prayer retreat (1.5 - 3 hours)

Adoration

Some people enjoy coming into God's presence through music. Sometimes, I pull out my guitar or listen to YouTube videos. One could also read a prayer or moving bible passage. I try to listen to the Holy Spirit and what songs or passages God is putting in my heart to draw near to Him. Toward the end of the time, read Psalm 19. Consider how God has impacted your life. Consider the greatness of God. Feel free to write any thoughts in the space below.

_____

_____

_____

_____

_____

_____

_____

_____

_____

_____

Confession

Read Romans 10, focusing on verse 9. Seek the Lord. What needs to be confessed before Him? Attitudes, thoughts, bad language? Read through the 10 commandments. Cry aloud to the Lord.

_____

_____

_____

_____

_____

_____

_____

_____

_____

_____

_____

_____

_____

Thanksgiving

Take time to journal things you are thankful for. Write about your life, health, etc. Write about anything in the reflection sections you feel grateful for. Read Nehemiah 8, focusing on verse 10.

Intercession

Begin to pray for others. This section is key as we can often get self-focused. How can you pray for immediate family, extended family, pastors, co-workers, nations in need, your country, people in leadership, and non-profits? Let the words flow. Read Ephesians 1, focusing on verse 18.

_____

_____

_____

_____

_____

_____

_____

_____

_____

_____

_____

_____

Searching

Humbly begin praying for yourself. Make a list of needs and give them one at a time to God, expressing your heart. What events are coming up in your life? What career goals do you have? What life goals do you have? Do any words stick out from your confession time? God is listening. Read Job 38 or a passage from any biblical character that inspires you.

_____

_____

_____

_____

_____

_____

_____

_____

_____

_____

_____

_____

Listening

Begin listening to God for passages of scripture/answers to prayer. If nothing sticks out, feel free to use the reflection scriptures below.

Scriptures for reflection on Patience:

2 Corinthians 6:4-10

Proverbs 19:11

James 5:7

_____

_____

_____

_____

_____

_____

_____

_____

_____

_____

_____

Application

Pray for strength to carry out anything God has revealed to you during the prayer retreat today. Read Psalm 46.

Reflection (30 minutes - 1 hour) separate from prayer retreat

Past

1. What regrets do you have from your past?

_____

_____

_____

_____

_____

_____

_____

_____

_____

_____

_____

_____

_____

_____

_____

_____

_____

_____

2. What are your favorite childhood memories?

3. What past events might you consider therapy for?

_____

_____

_____

_____

_____

_____

_____

_____

_____

_____

_____

_____

_____

_____

_____

_____

_____

4. Where was your favorite place to vacation as a kid?

5. What awards/recognition did you receive as a child?

_____

_____

_____

_____

_____

_____

_____

_____

_____

_____

_____

_____

_____

_____

_____

_____

_____

Journaling (10 minutes)

What stuck out to you today?

_____

_____

_____

_____

_____

_____

_____

_____

_____

_____

_____

_____

_____

_____

Week 2, day 3: Kindness part one

Visit your schedule to re-familiarize yourself with the rhythm you have set up. An example can be found in the "Notes on Sabbatical" section. See how you may want to alter it for today.

Prayer retreat (1.5 - 3 hours)

Adoration

Some people enjoy coming into God's presence through music. Sometimes, I pull out my guitar or listen to YouTube videos. One could also read a prayer or moving bible passage. I try to listen to the Holy Spirit and what songs or passages God is putting in my heart to draw near to Him. Toward the end of the time, read Psalm 19. Consider how God has impacted your life. Consider the greatness of God. Feel free to write any thoughts in the space below.

_____

_____

_____

_____

_____

_____

_____

_____

_____

_____

Confession

Read Romans 10, focusing on verse 9. Seek the Lord. What needs to be confessed before Him? Attitudes, thoughts, bad language? Read through the 10 commandments. Cry aloud to the Lord.

_____

_____

_____

_____

_____

_____

_____

_____

_____

_____

_____

_____

Thanksgiving

Take time to journal things you are thankful for. Write about your life, health, etc. Write about anything in the reflection sections you feel grateful for. Read Nehemiah 8, focusing on verse 10.

Intercession

Begin to pray for others. This section is key as we can often get self-focused. How can you pray for immediate family, extended family, pastors, co-workers, nations in need, your country, people in leadership, and non-profits? Let the words flow. Read Ephesians 1, focusing on verse 18.

_____

_____

_____

_____

_____

_____

_____

_____

_____

_____

_____

_____

_____

Humbly begin praying for yourself. Make a list of needs and give them one at a time to God, expressing your heart. What events are coming up in your life? What career goals do you have? What life goals do you have? Do any words stick out from your confession time? God is listening. Read Job 38 or a passage from any biblical character that inspires you.

_____

_____

_____

_____

_____

_____

_____

_____

_____

_____

_____

Listening

Begin listening to God for passages of scripture/answers to prayer. If nothing sticks out, feel free to use the reflection scriptures below.

Scriptures for reflection on Kindness:

Romans 2:4

Psalm 106:7

Application

Pray for strength to carry out anything God has revealed to you during the prayer retreat today. Read Psalm 46.

Reflection (30 minutes - 1 hour) separate from prayer retreat

Past

1. What sticks out to you from the last 10 years?

_____

_____

_____

_____

_____

_____

_____

_____

_____

_____

_____

_____

_____

2. What makes you cry from the past 10 years?

_____

_____

_____

_____

_____

_____

_____

_____

_____

_____

_____

_____

_____

_____

_____

_____

3. What from the past 10 years makes you laugh?

_____

_____

_____

_____

_____

_____

_____

_____

_____

_____

_____

_____

_____

_____

4. What embarrassed you in the past 10 years?

5. What was your favorite activity in the past 10 years? (Sports, games, etc.)

_____

_____

_____

_____

_____

_____

_____

_____

_____

_____

_____

_____

_____

_____

Journaling (10 minutes)

What stuck out to you today?

_____

_____

_____

_____

_____

_____

_____

_____

_____

_____

_____

_____

_____

Week 2, day 4: Kindness part two

Visit your schedule to re-familiarize yourself with the rhythm you have set up. An example can be found in the "Notes on Sabbatical" section. See how you may want to alter it for today.

Prayer retreat (1.5 - 3 hours)

Adoration

Some people enjoy coming into God's presence through music. Sometimes, I pull out my guitar or listen to YouTube videos. One could also read a prayer or moving bible passage. I try to listen to the Holy Spirit and what songs or passages God is putting in my heart to draw near to Him. Toward the end of the time, read Psalm 19. Consider how God has impacted your life. Consider the greatness of God. Feel free to write any thoughts in the space below.

_____

_____

_____

_____

_____

_____

_____

_____

_____

_____

Confession

Read Romans 10, focusing on verse 9. Seek the Lord. What needs to be confessed before Him? Attitudes, thoughts, bad language? Read through the 10 commandments. Cry aloud to the Lord.

_____

_____

_____

_____

_____

_____

_____

_____

_____

_____

_____

_____

_____

Thanksgiving

Take time to journal things you are thankful for. Write about your life, health, etc. Write about anything in the reflection sections you feel grateful for. Read Nehemiah 8, focusing on verse 10.

_____

_____

_____

_____

_____

_____

_____

_____

_____

_____

_____

_____

Begin to pray for others. This section is key as we can often get self-focused. How can you pray for immediate family, extended family, pastors, co-workers, nations in need, your country, people in leadership, and non-profits? Let the words flow. Read Ephesians 1, focusing on verse 18.

_____

_____

_____

_____

_____

_____

_____

_____

_____

_____

_____

_____

Humbly begin praying for yourself. Make a list of needs and give them one at a time to God, expressing your heart. What events are coming up in your life? What career goals do you have? What life goals do you have? Do any words stick out from your confession time? God is listening. Read Job 38 or a passage from any biblical character that inspires you.

Listening

Begin listening to God for passages of scripture/answers to prayer. If nothing sticks out, feel free to use the reflection scriptures below.

Scriptures for reflection on Kindness:

2 Samuel 22:51

Romans 11:22

_____

_____

_____

_____

_____

_____

_____

_____

_____

_____

_____

Pray for strength to carry out anything God has revealed to you during the prayer retreat today. Read Psalm 46.

Reflection (30 minutes - 1 hour) separate from prayer retreat

Future

1. Do you feel you are in the place God has placed you (why/why not)?

_____

_____

_____

_____

_____

_____

_____

_____

_____

_____

_____

_____

_____

2. Do you have a calling?

_____

_____

_____

_____

_____

_____

_____

_____

_____

_____

_____

_____

_____

_____

3. Do you have a purpose?

_____

_____

_____

_____

_____

_____

_____

_____

_____

_____

_____

_____

_____

_____

4. Are you fulfilling your calling/purpose?

5. What books do you like to read? (Why?)

Journaling (10 minutes)

What stuck out to you today?

_____

_____

_____

_____

_____

_____

_____

_____

_____

_____

_____

_____

_____

Week 2, day 5: Goodness

Visit your schedule to re-familiarize yourself with the rhythm you have set up. An example can be found in the "Notes on Sabbatical" section. See how you may want to alter it for today.

Prayer retreat (1.5 - 3 hours)

Adoration

Some people enjoy coming into God's presence through music. Sometimes, I pull out my guitar or listen to YouTube videos. One could also read a prayer or moving bible passage. I try to listen to the Holy Spirit and what songs or passages God is putting in my heart to draw near to Him. Toward the end of the time, read Psalm 19. Consider how God has impacted your life. Consider the greatness of God. Feel free to write any thoughts in the space below.

Confession

Read Romans 10, focusing on verse 9. Seek the Lord. What needs to be confessed before Him? Attitudes, thoughts, bad language? Read through the 10 commandments. Cry aloud to the Lord.

Thanksgiving

Take time to journal things you are thankful for. Write about your life, health, etc. Write about anything in the reflection sections you feel grateful for. Read Nehemiah 8, focusing on verse 10.

Begin to pray for others. This section is key as we can often get self-focused. How can you pray for immediate family, extended family, pastors, co-workers, nations in need, your country, people in leadership, and non-profits? Let the words flow. Read Ephesians 1, focusing on verse 18.

_____

_____

_____

_____

_____

_____

_____

_____

_____

_____

_____

Humbly begin praying for yourself. Make a list of needs and give them one at a time to God, expressing your heart. What events are coming up in your life? What career goals do you have? What life goals do you have? Do any words stick out from your confession time? God is listening. Read Job 38 or a passage from any biblical character that inspires you.

Listening

Begin listening to God for passages of scripture/answers to prayer. If nothing sticks out, feel free to use the reflection scriptures below.

Scriptures for reflection on Goodness:

Psalm 27:13-14

2 Thessalonians 1:11

Psalm 103:5a

Psalm 23:6

2 Peter 1:5

Ephesians 2:10

Pray for strength to carry out anything God has revealed to you during the prayer retreat today. Read Psalm 46.

Reflection (30 minutes - 1 hour) separate from prayer retreat

Future

1. What books have stuck out to you over the past 10 years?

_____

_____

_____

_____

_____

_____

_____

_____

_____

_____

_____

_____

_____

_____

2. What lessons have you learned in the past 10 years?

3. What can you take with you from the past 10 years?

_____

_____

_____

_____

_____

_____

_____

_____

_____

_____

_____

_____

_____

_____

_____

4. What are you willing to give up?

5. What hills are you willing to die on?

_____

_____

_____

_____

_____

_____

_____

_____

_____

_____

_____

_____

_____

_____

_____

Journaling (10 minutes)

What stuck out to you today?

_____

_____

_____

_____

_____

_____

_____

_____

_____

_____

_____

_____

Week 3, day 1: Faithfulness part one

Visit your schedule to re-familiarize yourself with the rhythm you have set up. An example can be found in the "Notes on Sabbatical" section. See how you may want to alter it for today.

Prayer retreat (1.5 - 3 hours)

Adoration

Some people enjoy coming into God's presence through music. Sometimes, I pull out my guitar or listen to YouTube videos. One could also read a prayer or moving bible passage. I try to listen to the Holy Spirit and what songs or passages He is putting on my heart to draw near to God. Toward the end of the time, read Psalm 19. Consider how God has impacted your life. Consider the greatness of God. Feel free to write any thoughts in the space below.

Read Romans 10, focusing on verse 9. Seek the Lord. What needs to be confessed before Him? Attitudes, thoughts, bad language? Read through the 10 commandments. Cry aloud to the Lord.

_____

_____

_____

_____

_____

_____

_____

_____

_____

_____

_____

_____

Thanksgiving

Take time to journal things you are thankful for. Write about your life, health, etc. Write about anything in the reflection sections you feel grateful for. Read Nehemiah 8, focusing on verse 10.

Intercession

Begin to pray for others. This section is key as we can often get self-focused. How can you pray for immediate family, extended family, pastors, co-workers, nations in need, your country, people in leadership, and non-profits? Let the words flow. Read Ephesians 1, focusing on verse 18.

_____

_____

_____

_____

_____

_____

_____

_____

_____

_____

_____

Humbly begin praying for yourself. Make a list of needs and give them one at a time to God, expressing your heart. What events are coming up in your life? What career goals do you have? What life goals do you have? Do any words stick out from your confession time? God is listening. Read Job 38 or a passage from any biblical character that inspires you.

_____

_____

_____

_____

_____

_____

_____

_____

_____

_____

_____

Listening

Begin listening to God for passages of scripture/answers to prayer. If nothing sticks out, feel free to use the reflection scriptures below.

Scripture for reflection on Faithfulness:

Psalm 18:25-26

_____

_____

_____

_____

_____

_____

_____

_____

_____

_____

_____

_____

Pray for strength to carry out anything God has revealed to you during the prayer retreat today. Read Psalm 46.

Reflection (30 minutes - 1 hour)

Adventure

1. Where have you traveled in the past 10 years?

_____

_____

_____

_____

_____

_____

_____

_____

_____

_____

_____

_____

_____

2. What sporting events have you attended in the past 10 years?

_____

_____

_____

_____

_____

_____

_____

_____

_____

_____

_____

_____

_____

_____

_____

3. What plays/movies/theatre have you been to see?

_____

_____

_____

_____

_____

_____

_____

_____

_____

_____

_____

_____

_____

_____

_____

4. What is your favorite movie?

5. What is your favorite song?

_____

_____

_____

_____

_____

_____

_____

_____

_____

_____

_____

_____

_____

_____

_____

_____

Journaling (10 minutes)

What stuck out to you today?

_____

_____

_____

_____

_____

_____

_____

_____

_____

_____

_____

_____

_____

_____

Week 3, day 2: Faithfulness part two

Visit your schedule to re-familiarize yourself with the rhythm you have set up. An example can be found in the "Notes on Sabbatical" section. See how you may want to alter it for today.

Prayer retreat (1.5 - 3 hours) separate from prayer retreat

Adoration

Some people enjoy coming into God's presence through music. Sometimes, I pull out my guitar or listen to YouTube videos. One could also read a prayer or moving bible passage. I try to listen to the Holy Spirit and what songs or passages God is putting in my heart to draw near to Him. Toward the end of the time, read Psalm 19. Consider how God has impacted your life. Consider the greatness of God. Feel free to write any thoughts in the space below.

Confession

Read Romans 10, focusing on verse 9. Seek the Lord. What needs to be confessed before Him? Attitudes, thoughts, bad language? Read through the 10 commandments. Cry aloud to the Lord.

Thanksgiving

Take time to journal things you are thankful for. Write about your life, health, etc. Write about anything in the reflection sections you feel grateful for. Read Nehemiah 8, focusing on verse 10.

Begin to pray for others. This section is key as we can often get self-focused. How can you pray for immediate family, extended family, pastors, co-workers, nations in need, your country, people in leadership, and non-profits? Let the words flow. Read Ephesians 1, focusing on verse 18.

_____

_____

_____

_____

_____

_____

_____

_____

_____

_____

Searching

Humbly begin praying for yourself. Make a list of needs and give them one at a time to God, expressing your heart. What events are coming up in your life? What career goals do you have? What life goals do you have? Do any words stick out from your confession time? God is listening. Read Job 38 or a passage from any biblical character that inspires you.

Listening

Begin listening to God for passages of scripture/answers to prayer. If nothing sticks out, feel free to use the reflection scriptures below.

Scripture for reflection on Faithfulness:

Luke 18:1-8

_____

_____

_____

_____

_____

_____

_____

_____

_____

_____

_____

_____

_____

Pray for strength to carry out anything God has revealed to you during the prayer retreat today. Read Psalm 46.

Reflection (30 minutes - 1 hour) separate from prayer retreat

Adventure

1. What creative thing have you always wanted to do?

_____

_____

_____

_____

_____

_____

_____

_____

_____

_____

_____

_____

_____

_____

2. What are you grateful for?

3. Beach, Mountains, or...?

4. What is the biggest adrenaline rush you have felt in the past 10 years?

_____

_____

_____

_____

_____

_____

_____

_____

_____

_____

_____

_____

_____

_____

5. Where are you most calm?

_____

_____

_____

_____

_____

_____

_____

_____

_____

_____

_____

_____

_____

Journaling (10 minutes)

What stuck out to you today?

_____

_____

_____

_____

_____

_____

_____

_____

_____

_____

_____

_____

_____

Week 3, day 3: Gentleness

Visit your schedule to re-familiarize yourself with the rhythm you have set up. An example can be found in the "Notes on Sabbatical" section. See how you may want to alter it for today.

Prayer retreat (1.5 - 3 hours)

Adoration

Some people enjoy coming into God's presence through music. Sometimes, I pull out my guitar or listen to YouTube videos. One could also read a prayer or moving bible passage. I try to listen to the Holy Spirit and what songs or passages God is putting in my heart to draw near to Him. Toward the end of the time, read Psalm 19. Consider how God has impacted your life. Consider the greatness of God. Feel free to write any thoughts in the space below.

_____

_____

_____

_____

_____

_____

_____

_____

_____

Read Romans 10, focusing on verse 9. Seek the Lord. What needs to be confessed before Him? Attitudes, thoughts, bad language? Read through the 10 commandments. Cry aloud to the Lord.

_____

_____

_____

_____

_____

_____

_____

_____

_____

_____

_____

_____

Thanksgiving

Take time to journal things you are thankful for. Write about your life, health, etc. Write about anything in the reflection sections you feel grateful for. Read Nehemiah 8, focusing on verse 10.

Intercession

Begin to pray for others. This section is key as we can often get self-focused. How can you pray for immediate family, extended family, pastors, co-workers, nations in need, your country, people in leadership, and non-profits? Let the words flow. Read Ephesians 1, focusing on verse 18.

_____

_____

_____

_____

_____

_____

_____

_____

_____

_____

_____

Searching

Humbly begin praying for yourself. Make a list of needs and give
them one at a time to God, expressing your heart. What events
are coming up in your life? What career goals do you have? What
life goals do you have? Do any words stick out from your
confession time? God is listening. Read Job 38 or a passage from
any biblical character that inspires you.

Begin listening to God for passages of scripture/answers to prayer. If nothing sticks out, feel free to use the reflection scriptures below.

Scripture for reflection on Gentleness:

Philippians 4:5

Colossians 3:12

Ephesians 4:26

Hebrews 12:14-15

Application

Pray for strength to carry out anything God has revealed to you during the prayer retreat today. Read Psalm 46.

Reflection (30 minutes - 1 hour) separate from prayer retreat

Habits

1. What habits are good for your career?

_____

_____

_____

_____

_____

_____

_____

_____

_____

_____

_____

_____

_____

2. What habits are good for your family?

_____

_____

_____

_____

_____

_____

_____

_____

_____

_____

_____

_____

_____

_____

3. What are some of your good/bad habits?

_____

_____

_____

_____

_____

_____

_____

_____

_____

_____

_____

_____

_____

_____

4. What are some habits you would like to break?

_____

_____

_____

_____

_____

_____

_____

_____

_____

_____

_____

_____

_____

_____

_____

A. What plan do you have for breaking these habits?

_____

_____

_____

_____

_____

_____

_____

_____

_____

_____

_____

_____

_____

_____

_____

Journaling (10 minutes)

What stuck out to you today?

_____

_____

_____

_____

_____

_____

_____

_____

_____

_____

_____

_____

_____

_____

Week 3, day 4: Self-Control part one

Visit your schedule to re-familiarize yourself with the rhythm you have set up. An example can be found in the "Notes on Sabbatical" section. See how you may want to alter it for today.

Prayer retreat (1.5 - 3 hours)

Some people enjoy coming into God's presence through music. Sometimes, I pull out my guitar or listen to YouTube videos. One could also read a prayer or moving bible passage. I try to listen to the Holy Spirit and what songs or passages God is putting in my heart to draw near to Him. Toward the end of the time, read Psalm 19. Consider how God has impacted your life. Consider the greatness of God. Feel free to write any thoughts in the space below.

Confession

Read Romans 10, focusing on verse 9. Seek the Lord. What needs to be confessed before Him? Attitudes, thoughts, bad language? Read through the 10 commandments. Cry aloud to the Lord.

_____

_____

_____

_____

_____

_____

_____

_____

_____

_____

_____

Thanksgiving

Take time to journal things you are thankful for. Write about your life, health, etc. Write about anything in the reflection sections you feel grateful for. Read Nehemiah 8, focusing on verse 10.

Begin to pray for others. This section is key as we can often get self-focused. How can you pray for immediate family, extended family, pastors, co-workers, nations in need, your country, people in leadership, and non-profits? Let the words flow. Read Ephesians 1, focusing on verse 18.

_____

_____

_____

_____

_____

_____

_____

_____

_____

_____

_____

Searching

Humbly begin praying for yourself. Make a list of needs and give them one at a time to God, expressing your heart. What events are coming up in your life? What career goals do you have? What life goals do you have? Do any words stick out from your confession time? God is listening. Read Job 38 or a passage from any biblical character that inspires you.

Listening

Begin listening to God for passages of scripture/answers to prayer. If nothing sticks out, feel free to use the reflection scriptures below.

Scripture for reflection on Self-Control:

Proverbs 16:32

Proverbs 25:28

James 3:4-5

Psalms 19:4

_____

_____

_____

_____

_____

_____

_____

_____

_____

Application

Pray for strength to carry out anything God has revealed to you during the prayer retreat today. Read Psalm 46.

Reflection (30 minutes - 1 hour) separate from prayer retreat

Relationships

1. What relationships in your life are healthy?

_____

_____

_____

_____

_____

_____

_____

_____

_____

_____

_____

_____

_____

_____

2. What relationships in your life are toxic?

_____

_____

_____

_____

_____

_____

_____

_____

_____

_____

_____

_____

_____

_____

_____

_____

_____

3. Do you consider getting a pet?

4. Do you want kids?

5. Have you considered adoption?

_____

_____

_____

_____

_____

_____

_____

_____

_____

_____

_____

_____

_____

_____

Journaling (10 minutes)

What stuck out to you today?

_____

_____

_____

_____

_____

_____

_____

_____

_____

_____

_____

_____

_____

_____

Week 3, day 5: Self-Control part two

Visit your schedule to re-familiarize yourself with the rhythm you have set up. An example can be found in the "Notes on Sabbatical" section. See how you may want to alter it for today.

Prayer retreat (1.5 - 3 hours)

Adoration

Some people enjoy coming into God's presence through music. Sometimes, I pull out my guitar or listen to YouTube videos. One could also read a prayer or moving bible passage. I try to listen to the Holy Spirit and what songs or passages God is putting in my heart to draw near to Him. Toward the end of the time, read Psalm 19. Consider how God has impacted your life. Consider the greatness of God. Feel free to write any thoughts in the space below.

Confession

Read Romans 10, focusing on verse 9. Seek the Lord. What needs to be confessed before Him? Attitudes, thoughts, bad language? Read through the 10 commandments. Cry aloud to the Lord.

_____

_____

_____

_____

_____

_____

_____

_____

_____

_____

_____

_____

Thanksgiving

Take time to journal things you are thankful for. Write about your life, health, etc. Write about anything in the reflection sections you feel grateful for. Read Nehemiah 8, focusing on verse 10.

Intercession

Begin to pray for others. This section is key as we can often get self-focused. How can you pray for immediate family, extended family, pastors, co-workers, nations in need, your country, people in leadership, and non-profits? Let the words flow. Read Ephesians 1, focusing on verse 18.

_____

_____

_____

_____

_____

_____

_____

_____

_____

_____

Searching

Humbly begin praying for yourself. Make a list of needs and give them one at a time to God, expressing your heart. What events are coming up in your life? What career goals do you have? What life goals do you have? Do any words stick out from your confession time? God is listening. Read Job 38 or a passage from any biblical character that inspires you.

Listening

Begin listening to God for passages of scripture/answers to prayer. If nothing sticks out, feel free to use the reflection scriptures below.

Scripture for reflection on Self-Control:

Titus 2

_____

_____

_____

_____

_____

_____

_____

_____

_____

_____

_____

_____

Pray for strength to carry out anything God has revealed to you during the prayer retreat today. Read Psalm 46.

Reflection (30 minutes - 1 hour) separate from prayer retreat

Relationships

1. How can you change to better a relationship?

_____

_____

_____

_____

_____

_____

_____

_____

_____

_____

_____

_____

_____

2. Do you need therapy for any of your relationships?

_____

_____

_____

_____

_____

_____

_____

_____

_____

_____

_____

_____

_____

_____

_____

3. What conversations have stuck out to you in the past 10 years?

4. What relationships would you like to work on?

5. What relationships are you praying for?

_____

_____

_____

_____

_____

_____

_____

_____

_____

_____

_____

_____

_____

_____

_____

Journaling (10 minutes)

What stuck out to you today?

Further Remarks

This next section covers some last thoughts for the sabbatical. Feel free to reference them as needed.

Therapy

Many of the younger generations embrace therapy. I believe we can give therapy to each other. Whether from a Grandparent, Aunt, Church Friend, Colleague, or professional Therapist, find someone to reach out to if any problem areas are brought to light during your sabbatical.

Prayer Retreat

Prayer retreat can be different for different people. I advise finding a quiet place where you can listen to music and focus. I have done prayer retreats in airports, hotel rooms, retreat centers and in my own living room. Be creative.

Reflection Questions

There are literally thousands of reflection questions on the internet. I have been blessed with many sources of reflection over the years and was able to come up with the reflection questions in this guide on my own. Be creative! Look up questions applicable to you or use the ones provided.

Conclusion

I believe every good manuscript needs a conclusion. My only advice to you is to be "Spirit-led" and let the Lord do the work. In the words of the passing of the peace, "The Lord bless you and keep you, the Lord make His face shine upon you and give you peace."

Job 11:13-15:

"13 Yet if you devote your heart to him

and stretch out your hands to him,

14 if you put away the sin that is in your hand

and allow no evil to dwell in your tent,

15 then you will lift up your face without shame;

you will stand firm and without fear." [1]

Bibliography

[1] *Holy Bible,* New International Version*,* Grand Rapids, MI, USA: Zondervan, 2002.

[2] Walt Disney Studios, Burbank, CA, USA. *Frozen 2.*

(Nov 22, 2019). Accessed:October 30, 2024.

[Online Video]. Available: https://www.disneyplus.com/

[3] L. Demetrician, LMFT, "Sabbatical Preparation", in *Desperate Rest: Restoring Your Soul Through Sabbatical,* 1st ed. Las Vagas, NV, USA: Independently Published, 2018, ch.4, pp. 76-82.

[4] Living Proof Ministries, Houston, TX, USA. *LIT Digital Bible Study.* (2016). Accessed: October 30, 2024. [Online Video].Available: https://store.lproof.org/

[5] J. Eldredge, "The Playfulness of God and the Poison of Religion", in *Beautiful Outlaw Experiencing the Playful, Disruptive, Extravagant Personality of Jesus*. New York, NY, USA: Faith Words, 2011, ch.1, pp.1.

[6] M. Balmer. (2018). The Church Jesus Designed Part 3 [Online]. Available: https://www.calvaryccm.com/melbourne/teachings/sermon/1213/the-church-jesus-designed-part-3

[7] Author Unknown. "Sacrament". Wikipedia.org. Accessed: October 30, 2024. [Online.] Available: https://en.wikipedia.org/wiki/Sacrament

[8] J. Eldredge, "Get Outside", in *Get Your Life Back: Everyday Practices For a World Gone Mad.* Nashville, TN, USA: Nelson Books, 2020, ch. 7, pp. 86.

[9] H. Blackaby, R. Blackaby and C. King, "Seven Realities of Experiencing God", in *Experiencing God: Knowing and Doing the Will of God, 2021 ed.* Nashville, TN, USA: B&H Publishing Group, 2021, ch.5, pp 62.

www.ingramcontent.com/pod-product-compliance
Lightning Source LLC
Chambersburg PA
CBHW041159220326
41597CB00001BA/4